Too Cute!
Baby Ducks

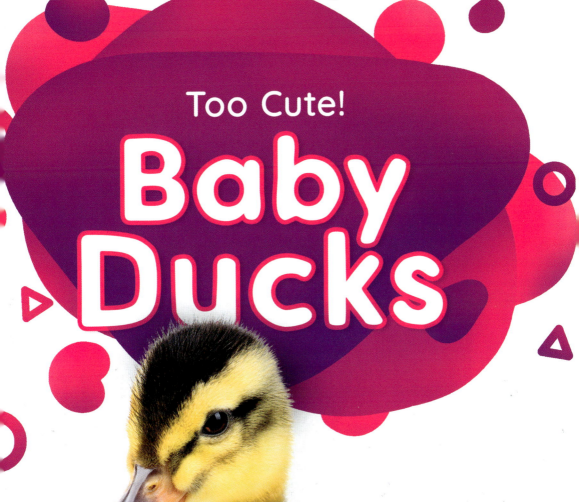

by Betsy Rathburn

BELLWETHER MEDIA • MINNEAPOLIS, MN

 **Blastoff! Beginners** are developed by literacy experts and educators to meet the needs of early readers. These engaging informational texts support young children as they begin reading about their world. Through simple language and high frequency words paired with crisp, colorful photos, Blastoff! Beginners launch young readers into the universe of independent reading.

Sight Words in This Book

a	from	the	two
and	get	their	water
are	go	them	with
at	in	they	
big	into	time	
eat	look	to	

This edition first published in 2022 by Bellwether Media, Inc.

No part of this publication may be reproduced in whole or in part without written permission of the publisher. For information regarding permission, write to Bellwether Media, Inc., Attention: Permissions Department, 6012 Blue Circle Drive, Minnetonka, MN 55343.

Library of Congress Cataloging-in-Publication Data

Names: Rathburn, Betsy, author.
Title: Baby ducks / Betsy Rathburn.
Description: Minneapolis, MN : Bellwether Media, 2022. | Series: Too cute! | Includes bibliographical references and index. | Audience: Ages 5-8 | Audience: Grades K-1
Identifiers: LCCN 2021040721 (print) | LCCN 2021040722 (ebook) | ISBN 9781644875728 (library binding) | ISBN 9781648345838 (ebook)
Subjects: LCSH: Ducklings--Juvenile literature.
Classification: LCC QL696.A52 R375 2022 (print) | LCC QL696.A52 (ebook) | DDC 598.4/11392--dc23
LC record available at https://lccn.loc.gov/2021040721
LC ebook record available at https://lccn.loc.gov/2021040722

Text copyright © 2022 by Bellwether Media, Inc. BLASTOFF! BEGINNERS and associated logos are trademarks and/or registered trademarks of Bellwether Media, Inc.

Editor: Amy McDonald   Designer: Jeffrey Kollock

Printed in the United States of America, North Mankato, MN.

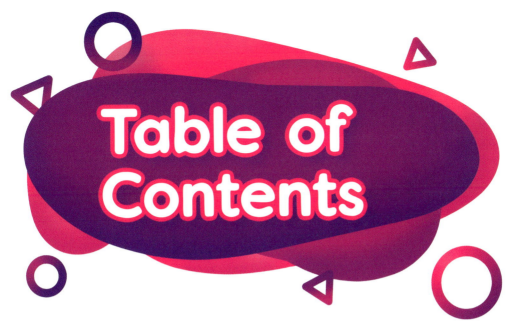

Table of Contents

A Baby Duck!	4
Fuzzy Followers	6
All Grown Up!	18
Baby Duck Facts	22
Glossary	23
To Learn More	24
Index	24

A Baby Duck!

Look at the baby duck.
Hello, duckling!

Fuzzy Followers

Ducklings **hatch** from eggs. They are fuzzy.

hatching

Ducklings follow mom.
They go in a line.

Ducklings hop into the water. They swim with **webbed feet**.

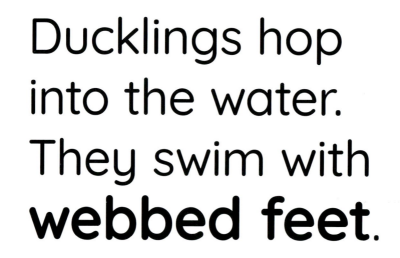

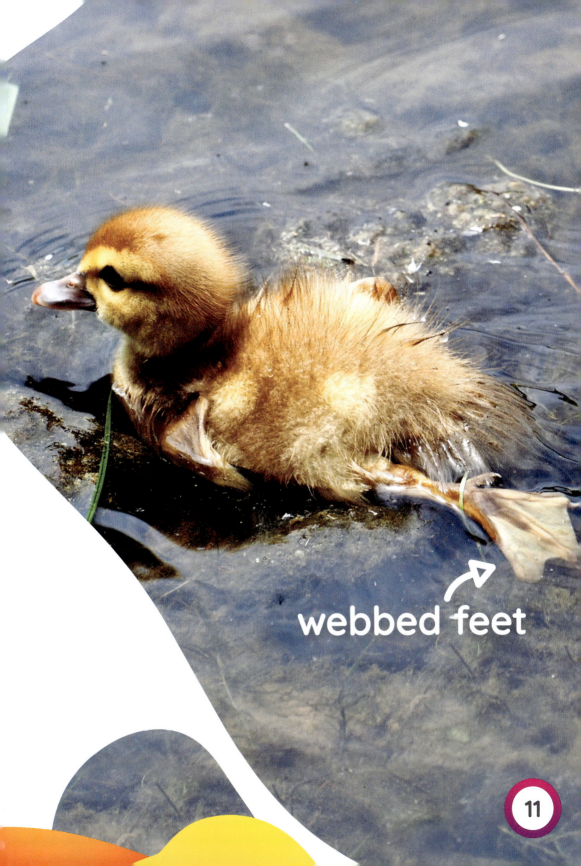

Ducklings eat with their **bills**. They grab bugs and plants.

Time to rest! Ducklings follow mom to the nest.

Ducklings sleep in the nest. Mom keeps them warm.

All Grown Up!

Ducklings grow fast. They get big **feathers**.

feathers

Ducklings leave mom after two months. Goodbye, ducklings!

Baby Duck Facts

Duck Life Stages

egg duckling adult

A Day in the Life

follow mom swim eat

Glossary

bills

the mouths of ducks

feathers

soft coverings that help keep ducks dry and warm

hatch

to break out of an egg

webbed feet

feet with thin skin that connects the toes

To Learn More

ON THE WEB

FACTSURFER

Factsurfer.com gives you a safe, fun way to find more information.

1. Go to www.factsurfer.com.
2. Enter "baby ducks" into the search box and click 🔍.
3. Select your book cover to see a list of related content.

Index

bills, 12, 13
bugs, 12
duck, 4
eat, 12
eggs, 6
feathers, 18
follow, 8, 14
grow, 18
hatch, 6

hop, 10
line, 8
mom, 8, 14, 16, 20
nest, 14, 16
plants, 12
sleep, 16
swim, 10
water, 10

webbed feet, 10, 11

The images in this book are reproduced through the courtesy of: DenisNata, front cover, pp. 5, 22 (duckling); iava777, p. 3; LeventeGyori, p. 4; Anneka, pp. 6, 23 (hatching); Silarock, pp. 6-7; Maria Oleinikova, pp. 8-9; VidEst, p. 10; Laura Burcham, pp. 10-11; Matias Gauthier, pp. 12-13; Kuttelvaserova Suchelova, p. 14; StoneMonkeyswk, pp. 14-15; Linda Ev, pp. 16-17; Ajax9, p. 18; Carlos Pereira M, pp. 18-19; Andi111, pp. 20-21; ncristian, p. 22 (egg); Aksenova Natalya, p. 22 (adult); CreativeMedia.org.uk, p. 22 (follow); Dark_Side, p. 22 (swim); Annette Shaff, p. 22; Dave Hansche, p. 23 (bills); Sketchart, p. 23 (feathers); Robin Keefe, p. 23 (webbed feet).